MEN-AT-ARMS SERIES

EDITOR: MARTIN WINDRO

Wellington's Light Cavalry

Text and colour plates by

BRYAN FOSTEN

OSPREY PUBLISHING LONDON

Published in 1982 by
Osprey Publishing Ltd
Member company of the George Philip Group
12–14 Long Acre, London WC2E 9LP

ISBN 0 85045 449 2

Filmset in Great Britain
Printed in Hong Kong

Introduction

Until 1745 the British cavalry comprised only 'Horse' and 'Medium' regiments; there were no light corps. However, during the Jacobite Rebellion the Duke of Kingston raised, at his own expense, a regiment of 'Light Dragoons' which, he proposed, would be similar to the hussar regiments already serving with success in other European armies. Although this unit was mounted on small light horses, and armed with curved swords and carbines suspended on swivels, there was little other concession to their 'light' status, and the uniform remained almost identical to that of the heavy dragoon regiments of the time. It was disbanded in 1746; but almost immediately another regiment was raised, largely from the same personnel, and the Duke of Cumberland assumed the colonelcy. Presently it achieved the status of a 'light' unit.

The regiment was disbanded in 1748, and the British Army had no light cavalry until April 1756 when, with some reluctance, Horse Guards agreed to the addition of a single 'light troop' to most cavalry regiments. These troops had an establishment of three officers, a troop quartermaster, five

Mameluke sabre of the 7th Hussars, *c.*1810. From the end of the 18th century light cavalry officers had begun to adopt an Oriental style of sabre known as the 'Mameluke'. They were often uniformly designed within a regiment, and had various types of scabbard and hilt. This example was used by Maj. Hodge, the officer who was wounded leading the charge against the French lancers at Genappe after Quatre Bras. (National Army Museum—as are all photographs in this book)

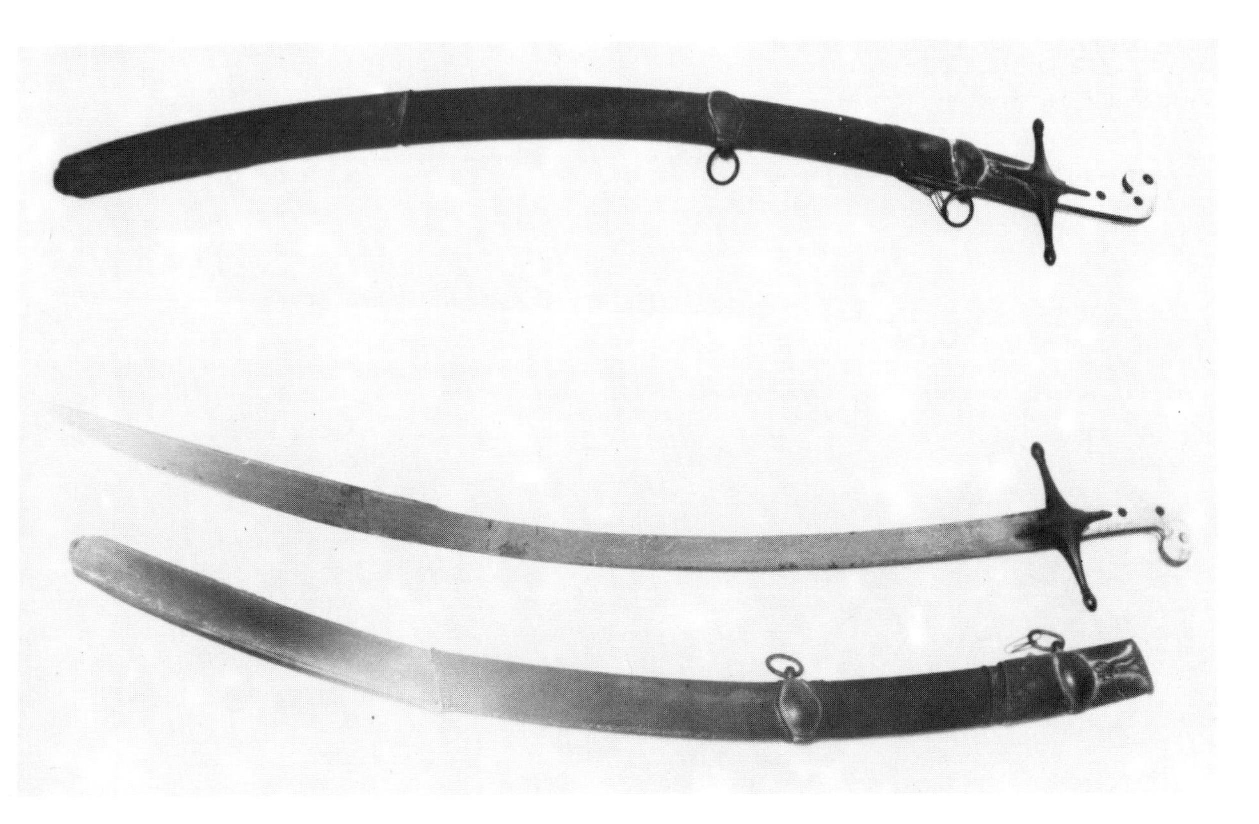

NCOs, two drummers and 60 'light and active' young troopers, and they were mounted on small, nimble horses. The light troops were dressed in the same uniforms as the remainder of their regiments except that they were given a new-pattern headdress—a small japanned leather helmet with an upright comb and front plate mounted with brass and a drooping feather. They were armed with carbines, pistols and straight swords without basket guards. From their formation these troops were fashionable and treated as having a special character, and they became so useful that by 1759 it was decided to form complete regiments of light cavalry.

The first was Elliot's, the 15th Light Dragoons, whose brilliant action in Germany earned them the honour 'Emsdorf', which was subsequently emblazoned on their helmets and colours. The 15th was followed by the 17th, 18th and 19th, and subsequently the 20th and 21st. In 1763, when hostilities ceased, the last two and the 17th were disbanded; the 18th became the 17th, and the 19th became the 18th.

Light Cavalry had now become an established part of the Army; and in 1776 the 8th and the 14th were transformed into Light Dragoons. The process continued, and in 1783 the 7th, 9th, 10th, 11th and 13th were similarly converted, the latter simultaneously changing their facings from green to buff as that colour was thought to be more appropriate to their new blue uniforms.

A 22nd Regiment was raised in 1760 but disbanded eight years later. A second 22nd was raised in 1779 but similarly disbanded in 1802; then a third regiment with the same number was

Ipswich Barracks, *c.*1808—a domestic scene, probably drawn by one of the Dighton family. It shows the back view of Lord Paget, on the left, and the Duke of Cumberland. Note the interior detail of the muff-like fur cap, the arrangement of lace on the back of the pelisse, and the large cocked hat on the table at the right.

formed from the 25th Light Dragoons, and was finally disbanded three years after Waterloo.

The 23rd Light Dragoons, raised in 1781 for special Indian service, became the 19th in 1783. A second 25th Regiment came into being in 1794 but was again disbanded in 1802, and the old 26th became the third regiment numbered the 23rd—only to be disbanded in 1817. There were two 24th Light Dragoons: the first was raised in 1794 and disbanded in 1802, and the second was formed from the 27th Light Dragoons and was not disbanded until 1818.

The 25th was raised as the 29th but was renumbered in 1802, although it is said to have retained its old number while serving with Lord Lake in India. It was finally disbanded in 1818. The 26th was raised in 1794 and renumbered the 23rd in 1802, and subsequently served with distinction in the Peninsula and at Waterloo. The 27th was raised in 1795 and was renumbered the 24th in 1802. The 28th existed between 1796 and 1802; and the 30th, 31st, 32nd and 33rd were all raised in 1794, but were disbanded only two years later.

The Horse Guards devised no specialised training manual for the Light Cavalry, but there was a lively awareness of this deficiency among the young officers of the various regiments, and in 1778 Captain Hinde published his '*Discipline of the Light Horse*' and dedicated it to the Colonel of the 12th Light Dragoons. It contains a mass of detailed information regarding the history, equipment and suggested training and duties of light cavalry which was far ahead of its time.

A trooper of the 7th Hussars—a plate by C. Hamilton Smith which gives a good overall representation of the hussar uniform. The cap is brown, its bag red, and the cap lines yellow. Note the simple trefoil ornament over the white cuff of the jacket, the white fur on the pelisse, and the double white stripe up the outward seam of the grey overalls. The shabraque is blue edged with white vandyking; and it is interesting to see that the 7th were already using sheepskins over the saddle.

In 1806 three of the regiments, the 7th, 10th and 15th, were remodelled as hussars; and the 18th was similarly converted in 1807. Nevertheless, although the transformation involved the introduction of much new, ornate and costly uniforms and equipment, the duties of the transformed regiments remained largely unaltered, and it was not until after Waterloo that the 9th, 12th and 16th Light Dragoons became lancers.

Light Cavalry Establishments

In 1800 the establishment of a regiment was ten troops, each having a strength of 90 rank and file. Two troops provided the depot squadron and the remaining eight formed four active or service squadrons. The troops were lettered—A, B, C, D and so on—and the squadrons were numbered, although the 1st was usually referred to as the Right Squadron when on parade. T. H. McGuffie, in an article titled '*The 7th Hussars in 1813*' (Journal of the Society for Army Historical Research, Volume XLII), records that when the regiment embarked for the Peninsula it was organised in squadrons, each of two troops, and that they were commanded by: *The Right*—The second lieutenant-colonel; *The 2nd*—The senior of the two troop commanders; *The 3rd*—The senior of the two troop commanders and *The 4th*—A Major. Apparently the Depot or Holding Squadron was also commanded by a major.

By 1813 the establishment of the light cavalry regiments was increased by two additional troops.

In May 1815 the effective strength of the cavalry of the King's German Legion provides a clear picture of the establishment of the regiments, and underlines the shortage of horses in the field:

1st Light Dragoons KGL

61 officers, 39 troop quartermasters and sergeants, 10 trumpeters, 958 rank and file—501 troop horses.

2nd Light Dragoons KGL

39 officers, 55 troop quartermasters and sergeants, 10 trumpeters, 498 rank and file—514 troop horses.

1st Hussars KGL

42 officers, 42 troop quartermasters and sergeants, 10 trumpeters, 518 rank and file—504 troop horses.

2nd Hussars KGL

41 officers, 55 troop quartermasters and sergeants, 10 trumpeters, 618 rank and file—675 troop horses.

3rd Hussars KGL

41 officers, 54 troop quartermasters and sergeants, 10 trumpeters, 564 rank and file—675 troop horses.

The following extracts from the last returns made by regiments in Belgium before the commencement of the Waterloo campaign provide an analysis of the establishments at that time:

An officer of the 10th Hussars, *c.*1811. This portrait of Lt. A. Finucane reveals no frame around the silver braid looping. The pelisse fur is grey, the lining crimson. Note plain pouch belt—this, and the sword belt, are both yellow faced with silver lace. The fur cap has a red bag and crimson and gold lines; the sword is apparently the 1796 pattern light cavalry sabre. Note clear detail of cuff lace.

A rather primitive, but nonetheless interesting portrait of an officer named Brunton of the 25th Light Dragoons, *c.*1812, showing the grey clothing used in the East Indies and other tropical stations. Note the barrelled sash, the cross belt with a plate on the breast, the facing-colour turban and chin scales to the helmet, and the wings on the jacket. The horse has a light-coloured canvas or web headstall.

Officer of the 18th Hussars, *c.*1812; note the extravagant plume socket and large aigrette feather in the fur cap; the development of the style and pattern of the silver braid looping on the uniform, and the belts. The sword is of Mameluke type, and the officer has laced pantaloons.

7th Hussars
1 lieutenant-colonel, 2 majors, 8 captains, 13 lieutenants, 1 paymaster, 1 lieutenant and adjutant, 1 quartermaster, 1 surgeon, 2 assistant surgeons, 1 veterinary surgeon.
10th Hussars
1 lieutenant-colonel, 1 major, 7 captains, 11 lieutenants, 1 paymaster, 1 lieutenant and adjutant, 1 assistant surgeon, 1 veterinary surgeon.
15th Hussars
1 lieutenant-colonel, 1 major, 8 captains, 10 lieutenants, 1 paymaster, 1 lieutenant and adjutant, 1 surgeon, 1 assistant surgeon, 1 veterinary surgeon.
18th Hussars
1 lieutenant-colonel, 6 captains, 16 lieutenants, 1 paymaster, 1 lieutenant and adjutant, 1 surgeon, 2 assistant surgeons, 1 veterinary surgeon.
11th Light Dragoons
1 lieutenant-colonel, 1 major, 7 captains, 9 lieutenants, 5 cornets, 1 paymaster, 1 adjutant, 1 quartermaster, 1 surgeon, 1 assistant surgeon.
16th Light Dragoons
1 lieutenant-colonel, 2 majors, 7 captains, 12 lieutenants, 3 cornets, 1 paymaster, 1 adjutant, 1 quartermaster, 1 surgeon, 1 assistant surgeon, 1 veterinary surgeon.
(*NB* the variations in these figures with those given by Siborne and listed under the Waterloo Order of Battle.)

Commissions

With few exceptions commissions were obtained by purchase. The prices fixed for cavalry regiments for the period of the Napoleonic Wars were as follows: lieutenant-colonel, £4,982 10s.; major, £3,882 10s.; captain, £2,782 10s.; and lieutenant, £997 10s. The price of 'a pair of colours' for a young gentleman in a fashionable cavalry regiment was thus very high, especially taking into account the correspondingly high cost of his elaborate uniform and accoutrements. When the regiment was in barracks or cantonments the officers formed messes, where the charges could also be very high. On service abroad they were entitled to draw the same rations as their men, although most supplemented the basic with additional wines and more expensive eatables when they were available.

After being commissioned there were three ways in which the officer could obtain promotion. He could either purchase a captaincy or majority; wait for his seniority in the Army, rather than in the regiment, to come to fruition; or depend on patronage. Consequently, officers tended to be from upper middle class and wealthy backgrounds, and in many cases had little aptitude for military careers. Under these circumstances it is remarkable that so many soldiers of merit emerged from such a system.

When, in 1810, the rank of troop quartermaster was abolished and replaced by troop sergeant major (troop corporal major in the Household Cavalry), a single regimental quartermaster was added to each regimental headquarters. These were commissioned officers, promoted from the ranks, and their commissions could not be pur-

Officers of the 12th Light Dragoons, 1794: a painting by Northcote of Pope Pius VI presenting medals and blessing officers of the regiment, painted some time after the event and thus showing later uniforms. These officers wear the more substantially braided jacket which succeeded the underjacket and shell of the 1790s. The picture provides a clear view of the handsome helmets, with the distinctive badges worn by this regiment.

Troopers of the 16th Light Dragoons, 1795, painted by W. Wollen; this provides an early glimpse of Light Dragoons wearing overalls and sheepskin saddle covers at the same time as the Tarleton helmet. Note the sabretasches and the bridles, which are already developing the style of the continental hussar.

chased. During the Napoleonic Wars many NCOs were granted commissions for exceptional deeds of gallantry in the field; for example, Sergeant Ewart of the Greys, who was given an Ensigncy in the 3rd Royal Veteran Battalion in recognition of his exploit at Waterloo.

The Light Cavalry Establishment, 1809

Regiment	*Establishment*	*Station*
7th Hussars	905	Home
8th Light Dragoons	720	East Indies
9th LD	905	Home, but Walcheren
10th H	905	Home
11th LD	905	Home
12th LD	905	Home, but Walcheren
13th LD	905	Home
14th LD	905	Peninsula
15th H	905	Home
16th LD	905	Peninsula
17th LD	940	East Indies
18th H	905	Home
19th LD	905	Home
20th LD	905	$\frac{1}{2}$ Sicily $\frac{1}{2}$ Peninsula
21st LD	905	Cape
22nd LD	928	East Indies
23rd LD	905	Peninsula
24th LD	928	East Indies
25th LD	940	East Indies
1st Hussars, KGL	694	Peninsula (June)
2nd Hs, KGL	694	Home, but Walcheren
3rd Hs, KGL	694	Home after Corunna

Officer of the 9th Light Dragoons in Review Order, 1813—a Hamilton Smith plate showing the pattern and style of the bridle and details of horse furniture. The shabraque and sabretasche are blue, edged with broad gold lace, and the black sheepskin has a red lining showing as 'wolves' teeth' around the edge. The review headstall is lined with red and the remainder of the bridle, reins, breaststrap, etc., all appear to be faced with gold lace. The crowns on the fore and hind parts of the shabraque and on the sabretasche are silver lined with red, and the Royal cyphers and numerals 'IX' also appear silver. Note that the pickers and chains are silver although the pouch belt itself is gold and crimson.

Drills, Tactics and Movements

nstructions to Light Dragoons regarding he exercise and field movements:

;ingle File: The front rank man marching singly ollowed by his rear rank man.

'ile: Two soldiers placed one behind the other vhen formed in ranks, but abreast when marching n file.

Rank: Two soldiers placed side by side.

;ections of Threes: Three men abreast, each rear ank three following its front rank.

;ub-Division: Half a Division.

)ivision: Fourth part of a squadron. Divisions vere numbered 1–2–3–4, from the right.

Troop: Half a squadron—known as 'right' or 'left' n each.

;quadron: Two troops—two or more compose he regiment—known as 1st, 2nd, 3rd etc, from he right.

3rigade: Two or more regiments of cavalry.

Regulated intervals between files in a squadron: ix inches from knee to knee.

Half open files: Eighteen inches from knee to nee.

)pen files: One yard from knee to knee.

From one troop to another in close column: One iorse length.

From one squadron to another in close column: Two horses length.

From one regiment to another in close column of quadrons: Four horses length.

From one regiment to another in close column of roops: Two horses length.

Regulated Pace

Walk: Not exceeding four miles an hour.

Trot: Not exceeding eight and half miles an hour.

Gallop: Eleven miles an hour.

Vedettes

Generally a single dragoon; at night and when the

)fficer of the 10th Hussars, *c.*1796—a good representation of he uniform, showing the flat lace frame around the looping. This officer appears to be wearing the underjacket without he shell, but note the chain wings on the shoulders. Note also he leopardskin turban, helmet badge, blue breeches with ilver lace, pouch belt with plate, and pouch on the hip; the abre is the 1796 pattern.

Corporal of the 10th Hussars, *c.*1812. The regiment discarded their old yellow facings and used scarlet for a short period, and Hamilton Smith records the fact in this plate. Note brown fur cap with red bag and yellow lines; white pelisse fur; and white looping and buttons. The barrelled sash is crimson and yellow. Above the left cuff of the jacket can be seen the extraordinary pattern of corporal's chevrons worn by the 10th, surmounted by the Prince of Wales's feathers and motto scroll. The red shabraque has white traced vandykes, double reversed and interlaced cyphers, and crowns. The valise is blue, the sabretasche plain black.

position was close to the enemy, two—one young soldier to be posted with a veteran trooper. By day vedettes were to take carbines but at night pistols were considered more convenient. They were instructed to take care not to fire and precipitate unnecessary action by the enemy. If a deserter or a flag of truce came in, the nearest picquet was signalled by raising the headdress on either a carbine or the sword. If a body of enemy troops approached, the vedette was instructed to put his horse to walking in a circle, increasing his pace proportionate to the speed and proximity of the enemy. His movement was to be repeated by all the other vedettes within sight. The signals would be picked up by the so-called 'warning sentry' posted mid-way between the outlying vedettes and the main picquets.

The vedette was given a note by the NCO in charge of the picquet reminding him that dust from a body of cavalry on the move rises higher than that from a body of infantry. The latter is lower and denser. Artillery and waggon movement causes uneven and unequal dust formation. At 2,000 yards he was told nothing could be discerned of infantry other than the glitter of the sun on their arms. At 1,500 yards he would be able to ascertain whether he was watching men on horseback. At 600 yards he would be able to count the number of men involved by heads; and by the time they were within 300–400 yards he would be able to pick out faces and uniforms. At 70–100 yards eyes could be seen as dots!

Swimming a river

Troops were to take water as a body as that helped to break the strength of currents of water. They were instructed to fasten their swords around their necks by the sword knots and hang them down their backs. Horses to be guided with the left hand, the right taking the mane in the centre of the neck, preserving a proper balance and sitting very steady in the saddle.

Attacking the enemy

The regiment would be brought up at a moderate pace, as near as possible to the enemy before being ordered to draw swords, and the order given to charge. It was considered a great error to draw swords before the very instant of attack, because (a) it was a caution to the antagonists and (b) it diminished the overall effect. As they drew swords they were ordered to strike spurs and fall upon the enemy, all as one and with the same impulse, and preferably on the flanks.

When attacking infantry

The instruction was to pace the charge so that the final impulsion was made immediately after they had fired, going through the smoke. Squares were attacked on the angles because this narrowed the amount of fire which could be brought to bear obliquely. It was emphasised that the moment

Trooper of the 13th Light Dragoons, *c*.1812—a cleverly designed Hamilton Smith plate which shows the turnbacks, cuffs, back seam piping on the sleeves, carbine swivel and pouch belt arrangements. The carbine is not attached to the swivel, and a close examination of the original plate does not show how it is meant to be suspended; note the lock cover. The shabraque is decorated at fore and hind parts with crowns over the GR cypher over 'XIII' over 'LD'. The circlet on the front of the cap is quite plain and has no numeral or device. In the background is a trooper in a cloak.

Officer of the 18th Hussars, *c*.1814. This portrait of an unknown officer shows the fur cap with its light blue bag and a rather simple aigrette. Note the plaited cap line, the olivette ends to the cords of the sash, and the Mameluke sabre. The jacket has a very simple white collar.

most likely to succeed was when the infantry formation was either executing a movement or when the formation had been shaken by the fire of artillery.

When charging artillery

On level ground the artillery were to be attacked independently, but where hollow ways, ravines or large troop formations intervened these were used to screen closer-order advances. In general it was found the better plan to attack artillery in two parties, the one feinting and threatening a false charge while the other made the real one. Infantry or cavalry supporting the artillery had also to be attacked to allow time for the gunners to

Cornet Fitzherbert Brooke, 26th Light Dragoons, 1811. This rather fine miniature shows clearly a variation in size between the centre row of buttons and those at the outer ends of the loops. Note also the odd lace arrangement on the collar.

be sabred and the drivers forced to bring the guns off. Where that was not possible the guns must be spiked, or if there were no nails for the purpose, the guns dismounted and all the horses taken or shot.

When supporting guns

It was considered best to place cavalry on their flank at a distance of about 100 yards, and concealed as well as possible. In such circumstances the commanding officer would place himself so as to observe as much as possible.

Pursuit

Pursuit was executed with as much rapidity as possible with an effort to secure as many prisoners as possible, but it was emphasised that it should be disciplined and orderly, and caution was to be observed to prevent the regiment falling into ambuscades and other dangers.

The March

Cavalry had to learn to ride across rough country, to jump when necessary, to keep in line when required and to keep their spacing while walking, cantering, trotting and galloping. During marche sections were detailed to act as screens, skir mishers, escorts, scouts and orderlies.

Marches over any considerable distance wer at all times, preferably, in columns of divisions c the line: that is, by half-squadrons, divisions, o sub-divisions when the squadrons exceeded 4 files. However, where the breadth of the rout required it, the march was made by a six-ma front (ranks of threes) or four-man front (ranks b twos), or alternatively, by a two-man front by th filing of ranks, although this was hardly ever done

During occasional halts some individual soldier were permitted to dismount, but only unde strict supervision. No man was permitted to rid in a careless, lounging manner, and no one ma was allowed to stop under the pretence of water ing his horse. At every general halt every necessar examination and adjustment of girths, saddles etc., had to be made.

The bat horses, servants with led horses an canteen horses remained with the squadrons Carriages and all the waggons followed and wer provided with an escort.

Colonel Thomas Hawker, 20th Light Dragoons, 1812. Th colonel is shown in the new Light Dragoon uniform. Th silver rank distinction is only vaguely discernible on the gol epaulette.

For the purposes of supply, the cavalry unit was the regiment. The allowance of transport was one mule for every two men and horses, so that a regiment of, say, 400 dragoons, with nearly 500 horses and baggage mules, required 300 commissariat mules to itself for the maintenance of food and fuel. For all the Army in the Peninsula 9,000–10,000 mules were needed over and above regimental transport.

Marches of cavalry regiments had to be so arranged that troops arrived at quarters no later than noon. The first and last miles of the march were always conducted at the walk, and occasional halts were arranged at the discretion of the officer commanding so as to allow girths to be examined and tightened where necessary, and to allow the men to make sure that their baggage was properly secured.

When they arrived at quarters troops had to forage in one hour and unsaddle, etc, in one and a half hours. When the troops were in possession of corn they carried it suspended from the saddle in corn sacks.

Any proposal to increase the pace of the march was done by means of a preparative order, which was not put into operation until the whole column had been so cautioned. When the walk was resumed after the trot it was done by means of a gradual decrease in pace.

When a regiment on the march met another, the youngest corps halted and was formed until the other corps had completely passed. Troops without standards paid compliments to troops with standards, whether senior or not. During the compliments standards were uncased, swords drawn and the officers saluted with all ceremony. Trumpeters of the passing regiment sounded the Parade March as they passed.

Troops marched in column in such fronts as the roads and circumstances allowed. The object was not to lengthen the line of march more than was necessary.

There was a so-called 'casual division' consisting of the forage carts, saddler's waggon, sutler's cart, light carriages, led horses and any dismounted men who were under the direction of an experienced NCO. It was usually marched as close to the rear of the main column as possible, but in some circumstances, where the officer commanding thought it expedient, it was sent on before.

Trooper of the 18th Hussars, *c.*1812—a soldierly figure, showing the hussar clothing of the period to perfection. The pale blue bag to the cap is just visible; note yellow cap lines, brass chin scales, and the pale brown bridles, etc. The dark blue shabraque has white vandykes; the valise is plain blue. The white sheepskin appears to have dark blue edging, and a blue cloak is rolled under the forepart of the shabraque just behind the pistol holster.

The rear guard comprised the 'rearguard' of the camp, the barrack guard (when the regiment was in barracks) or the daily guard (when in cantonments) and it marched about 100 yards in the rear of the main body, and of the casual division when that marched at the rear.

The regimental heavy baggage, with an escort proportionate to its size, generally drawn from each troop, could be ordered to march at an earlier hour, but this was left to the discretion of the officer commanding. Three waggons were allowed for the regimental staff, including their papers and the military stores. The adjutant had one waggon, and it included the paymaster's papers and other records. There was one waggon for the surgeon and veterinary officer and the assistants where appropriate; it carried their stores and instruments and the hospital stores. There was also a

Officer of the 14th Light Dragoons, 1812—a costume which seems to have been taken directly from the tailors' patterns of the period, although it seems unlikely that a gold-laced sabretasche would be worn with a uniform decorated with silver appointments. At Eton College a portrait of Lt.Col. Felton wearing the same uniform is preserved, but with a blue pelisse trimmed with black fur and black braid looping. In the field this uniform arrived very late, and Lt. Bretherton records that in 1814 he wore it with the bearskin-crested helmet.

regimental stores waggon in the charge of a clerk/storekeeper, a saddler's stores waggon, the armoury waggon and the forge waggons. For each troop there was an allowance of one and a half stores waggons; the regimental women travelled with them, selected by the quartermasters from a list, usually of the steadiest and most respectable, especially those who washed for the men. Each of these waggons carried about 30cwt—more in the case of the forges, although they were specially designed vehicles for the purpose, and only available to some regiments. Ninepence a mile was allowed for each field officer, ninepence a mile for a detachment with one or two subalterns, one shilling a mile for the captains, and threepence a mile for every 20 men of larger detachments than one troop, for lodging and forage. The allowance by regulation for the conveyance of baggage for each troop of one and a half waggons was one shilling and sixpence a mile.

Parades

The Troop

Formed into two ranks, the tallest men and horses in the front rank, sized from the right, except where the troop was to act in squadron, in which case they are to be on that flank which was to be the inward one. Should the numbers of men be uneven the last man but one on the outward flank of the front rank remained uncovered.

The troop was then 'told-off' into divisions, sub-divisions, threes and files. The 'telling-off' was from the centre; the senior subaltern was posted in front of the centre, the junior subaltern to the rear of it. NCOs were posted, one on each flank of the troop, two in the centre, the remainder as far as possible on the flanks of the sub-divisions. The trumpeter and the farrier were posted half a horse's length in the rear.

The Squadron

The squadron was 'told-off' in two troops of which the standard occupied the centre, then into divisions, threes and files, and when the squadron exceeded 64 files it was sub-divided into sub-divisions. In hussar regiments, which had no standard or guidon, the NCO on the right of the left troop was considered the centre.

The Divisions

The centre division was divided equally by threes.

Trooper of the 15th Hussars, *c.*1812—an interesting Hamilton Smith plate showing the reloading of the carbine while still attached to the swivel belt. The light brown fur cap has a red bag and yellow lines. The jacket has red facings, and braiding and buttons are white. The blue shabraque has solid red vandykes; the red devices are the Royal cypher reversed and interlaced below a scroll and crown and above two crossed lances, heads downward. The white sheepskin has a red edge, and the valise is plain blue.

Officer's helmet, 9th Light Dragoons, c.1800–1812. This is the so-called 'Tarleton' helmet-cap worn, with modifications, between about 1789 and the introduction of the shako in 1812. Note the sharply upswept front peak, the regimental name-plate above it, the pleated turban heavily guarded with treble chains, and the magnificent badge. The bearskin crest has unfortunately been flattened down: in pristine condition it would, of course, stand much higher. Later versions of the helmet had chin scales.

If the divisions were not all the same strength those on the flanks of the squadron to an extent not exceeding three files had to be the strongest. Thus, if the troops consisted of 20 files each, the centre divisions had to be nine and the flanks 11; if 21 files, the centre was nine and the flanks 12. However, if the troops were 22 files each, then the centre divisions were 12 and the flanks ten; if 23, the centre would be 12 and the flanks 11; and so on.

One sergeant carried the standard, in the centre of the front rank, covered by a corporal; one other, as a squadron marker, half a horse's length on the right of the squadron serrefile; the remainder on the flanks of all the divisions, all supernumeraries to be in the rear with the serrefiles.

Close Order

The squadron leader was posted half a horse's length in front of the standard; troop leaders the same distance in front of the centre of their troops; the squadron serrefiles in the rear of the centre of the squadron, and each troop serrefile in rear of the centre of each troop at the same distance. The trumpeters were stationed in the rear of the second file from each flank but in 'order', on the right of the front rank, at an interval of one horse's length.

The Squadron in Line

The farriers were placed so as to divide the interval between the troop serrefiles and the trumpeters.

NB: 'Telling-Off' was the operation by which a squadron or troop was divided in the manner necessary for its proper working in the field or on parade.

In the field

Advanced and Rear Guards

Their function was generally to keep communication with the other files and with the main body, of which they were never to lose sight. When the head of an advanced guard arrived at a thicket, or a turn in the road, or the top of the hill, especially in thick weather, one man was instructed to always keep the next man in succession in view so as to ensure that the party was kept in line for its right objective. The main body was similarly instructed in respect of the rear guard. At night tufts of straw as well as lamps, when available, were left to serve as markers for the route.

On coming to a hill, one man trotted up, another followed at the walk. The first advanced far enough to look about him, first removing his headdress. If he saw nothing he continued his

Sabretasche of the 9th Light Dragoons, 1815.

advance—that was the signal for the remainder of the guard to move on. If, on the other hand, he discovered the enemy, he reported immediately to the officer commanding his party, his place being taken by the next in file. If the enemy was close at hand and discovered suddenly the first guard would fire a shot.

The rear guard had the prime duty of making sure that no one remained behind. Any sick or wounded were placed in the waggons or put on spare horses which had been pressed for the purpose. If neither option was possible they were placed in the most comfortable and convenient position. The rear guard was answerable that no stores, ammunition or provisions fell into enemy hands. Any pursuit was to be impeded by destroying bridges, blocking and barricading gateways, setting up barricades and abatis across every lane or street, and obstructing all fords by throwing harrows, scythes, rakes and plough shares into the water.

Flanking Parties

These were ordered to explore the countryside on either side of the line of march, keeping observation on the movements of any enemy and covering the movements of any member of their own party. Their orders were to keep a sharp eye, to support each other, keeping in twos, so that while one fired or reconnoitred the other would always be loaded and ready to protect his colleague. When the regiment began to move into the attack flankers were called in. If they were not they gradually fell back and formed in at the rear, ensuring that the front remained unimpeded.

Skirmishers

Skirmishers were thrown out in advance of the regiment to 'feel' the way for the advance of troops to attack. Their duty was to harass and annoy the enemy, but at the same time to protect and conceal the movement of their main body. They were instructed to ensure that their saddles were well girthed up with short stirrups so that they could occasionally stand in them to improve observation. Holsters were exposed, the front of the shabraque was drawn back over the thighs, and in some circumstances it was considered expedient to draw sword, using the strap to secure it to the wrist. The carbine was primed and the weapon kept ready to hand.

Trooper of the 10th Light Dragoons, *c.*1803, in a neatly executed water colour by Richard Simpkin which shows a model light dragoon uniform of the period. Note the frame around the braid looping, the helmet badge, the white breeches, and the hussar-type boots.

Their line was always kept parallel to that of the enemy, and they had to observe carefully the nature of the ground occupied by the enemy in order that when working it themselves they would not be entirely at a loss. Whenever possible the skirmisher was instructed to gain the highest possible ground. If he observed enemy troops apparently preparing an ambush he was taught to observe them, using concealment, but to ensure that his officer was informed as soon as possible.

The skirmisher was selected for his coolness and intelligence. When he loaded he had to take his time, but to keep his horse in motion as much as possible to offer as poor a target as possible. To fire with any effect he had to halt and take deliberate aim, 'bringing the carbine firmly against the shoulder, then gradually raising the barrel until it was in line with the centre of the target . . .'. On no account was he to stray from his allotted position or to go after stray horses, baggage or any kind of plunder, nor to dismount without express permission. No skirmisher was allowed to go to the rear with a wound unless he was expressly ordered to do so.

After a March

Immediately the horse was got into stables or the bivouac the bridle was taken off, the crupper and

Officer of the 10th Light Dragoons, *c.*1809, in a water colour by Simpkin which shows the peakless shako used by the hussars in Spain. Note the cartwheel ornament on the front of the cap, the wearing of the pelisse with no frame around the looping, the overalls with buttoned fly and booting, and the red shabraque with silver traced vandyking. The bridle is decorated with cowrie shells.

breast plate were loosened, the valise, cloak, waterdeck and arms were taken off. The bit and the stirrups were wiped clean. The dragoon then shook a little loose litter under the horse, picked out his feet, turned him about and rubbed his head and ears with a dry bundle of straw called a 'wisp'. The dragoon then tied the horse up, 'wisped' him well under the belly and about the legs and gave him some fresh hay, making sure any old hay was taken out of the rack. Having completed those chores the dragoon was free to take his valise, cloak and arms to his quarters, put on his stable dress, and rub his arms over very carefully, making sure he did not use any abrasive material. When the trumpet sounded for 'stables' the horses were unsaddled, if they were cool, and their backs carefully examined for 'sores or warbles'. He checked that the shoes were fast and then proceeded to dress the horse.

Drills

The Foot Parades

For the manual exercise with the carbines the troops paraded with carbines and sidearms, and with sidearms only for the sword exercises. For marching and wheeling drills they paraded without weapons. These drills began about the middle of April and continued, as weather permitted, until the end of August. They were usually at 2pm daily. Field days, Saturdays and market days were exceptions for these parades, but not for general squad parades by the sergeants, which were held in the winter as well as the summer.

Marching Order Exercises

The troops were turned out in marching order at least once a month, generally on the second Friday when Saturday was a market day, but otherwise on the second Saturday, when each trooper had his clean linen from his washerwoman and had to have all his articles with him. The troops were exercised in walking and occasionally trotting along roads for one hour and sometimes two hours, depending on their efficiency.

Recruits: Riding and Foot Drills

These parades were carried out under the orders of the Officers Commanding and the Drill Sergeants whenever weather permitted. Such drills were arranged so as not to interfere with the men's dinner hour, and the riding drills so as not to interfere with the specified times for the feeding of the horses with oats.

Carbine Exercises on Foot

The men fell in with the carbines shouldered on the right side. The 'inspection motions' were then: 'port arms', 'open pans', 'shut pans', 'draw ramrods', 'return ramrods', 'shoulder arms', 'front'. 'Shouldering Arms from the swivel and springing' (*sic*) were also practised on foot.

On these parades the swords were suspended at the full length of the slings when the troops were halted. To enable them to move freely in drill circumstances they caught the sword up quickly in the left hand on the last word of all preparatives, or on the word 'march', after the word 'face' or the word 'wheel' or on the last word of 'wheel into

Officer of the 10th Hussars, *c.*1812. This handsome figure by Simpkin shows the uniform of the regiment at the time when red facings were worn. Note, again, the cowries on the bridle; the absence of the frame around the looping; and the pattern of the shabraque, which is the same as that shown in the earlier painting.

Trooper of the 10th Hussars—this painting by W. Wollen is attributed to 1815, but shows the uniform with the red facings which had been replaced by that date. Note solid vandyked border to the shabraque.

line'; and then dropped them again to the full extent of the slings, especially on the word 'halt' or 'halt front'. The same rule applied for the steadying of scabbards when swords were to be drawn.

Guard Duty

Guards were always mounted in proportion to the strength of the command. If three subalterns were available there would be an orderly officer. Where necessary a senior sergeant was appointed sergeant major; and if there was no orderly officer, the orderly sergeant was ordered to attend the assembly at the commanding officer's quarters and to make his report to him in the evening.

If it was an NCO's guard that was to be mounted the orderly officer coming off duty delivered to the CO the reports of the old guard on a form specially printed for the purpose. The orderly officer coming off duty inspected the old guard, and the sergeant of that guard then marched it off to the parade ground and dismissed it. The orderly officer coming on duty inspected the new guard, ordered it 'to beat', wheeled it into sections and ordered it to march off, when the NCO in charge took over. If it was an officers' guard the orderly officer had no charge of it nor did he report it.

Guards were not permitted to take off their clothes nor their accoutrements, nor were they allowed to sleep in their hats. After dark they took off their hats and replaced them with foraging caps. When the general watch was set and turned out, the guard turned out with them. The guard was turned out twice a day as a matter of routine, and always at the setting of the general watch, when patrols were sent out at night. The watch was set at 9pm, after which no soldier was permitted out of quarters without permission of the CO. All guards, posts and outlying picquets were turned out at daybreak.

By 9am the men of the old guard were ordered to have their hair freshly tied, their hats brushed, feathers adjusted, dress brushed and accoutrements clean. The orderly officer coming off duty had orders to confine any man immediately if he was dirty, or on parade in an un-soldierly manner.

Sentries were relieved every two hours—in very severe weather every hour. They always carried their carbines either supported or shouldered. Sentries were not permitted to talk to anyone, to smoke, to make impolite noises to passers-by or to whistle, but had to be attentive and to generally conduct themselves decently. They were not allowed to relieve themselves. The guard was not turned out for, nor did any sentry salute any officer unless he was in uniform. After 'retreat' no compliments were paid. Sentries cloaked in rainy weather paid no compliments but stood steady, turning the head as officers passed. If it was not raining at the time when the cloak was worn and an officer passed, the sentries threw back the cloak and saluted as usual. The guard was turned out to all general officers and to field officers of the regiment—to the former they presented and to the latter shouldered, except to the commanding officer, who was always complimented with the 'present'.

Sergeants had to examine arms and all ammunition before posting sentries and had to ensure that the flints were properly adjusted and screwed tight. At other times the carbines remained in the arms racks provided in the guard

1: Trooper, 10th (Prince of Wales's Own) Light Dragoons, 1787
2: Officer, 7th (Queen's Own) Light Dragoons, 1793
3: Trooper, 15th (or King's) Light Dragoons, 1794

1: Officer, 12th (the Prince of Wales's) Light Dragoons, 1805
2: Trooper, 11th Light Dragoons, 1798
3: Officer, 14th (Duchess of York's Own) Light Dragoons, 1800

B

1: Officer, 22nd Light Dragoons, 1807
2: Officer, 10th (or Prince of Wales's Own) Light Dragoons, 1805
3: Colonel, 7th (or Queen's Own) Light Dragoons, 1798

C

1: Trooper, 7th (Queen's Own) Light Dragoons, Hussars, 1808
2: Sergeant Major, 15th (or King's) Light Dragoons, Hussars, 1808
3: Officer, 10th (or Prince of Wales's) Hussars, 1808

1: Officer, 13th Light Dragoons, 1813
2: Sergeant, 20th Light Dragoons, 1812
3: Trooper, 22nd Light Dragoons, 1812

E

King's German Legion, 1815:
1: Officer, 1st Regiment of Light Dragoons
2: Officer, 2nd Regiment of Hussars
3: Trooper, 2nd Light Dragoon Regiment

F

1: Corporal, Staff Corps of Cavalry, 1814
2: Officer, 18th Hussars, 1814
3: Officer, 7th (Queen's Own) Light Dragoons, Hussars, 1810

1: Trooper, 7th (or Queen's Own) Light Dragoons, Hussars, 1815
2: Trooper, 18th Hussars, 1815
3: Officer, 10th (or Prince of Wales's Own) Hussars, 1815

H

room, according to each man's number in the guard.

The officer or NCO of the guard was responsible for seeing that the city, town or barrack gates were shut and locked at the ordered hour, and for keeping the keys on his person. The gate was thereafter only to be opened by the officer or sergeant of the guard.

No prisoner was to receive any food other than bread and water and no guard conversed with him. They had to be clean and had to dress their hair each morning. No liquor whatsoever was allowed in the guard room and no guard was permitted to absent himself to get drink. For that reason the officer or NCO was required to call the roll constantly. Food was brought to the guard in special tins otherwise kept in the troop store.

One of the principal guard duties was to be alert to stop any soldier becoming involved in a civil affray. They also turned out in case of fire. It was the adjutant's duty to make sure that the instructions of the guard were written and displayed in every quarter of the regiment, and the men were expected to make themselves acquainted with them before they were allocated for duty.

Light Dragoon Horses

Unlike the French hussars and chasseurs, who are referred to as riding small horses often no larger than ponies, the Inspection Reports reveal that the horses of British Light Dragoon and Hussar regiments differed only slightly from those ridden by the 'heavies'. For example, the report on the 10th Hussars for 1813 lists the sizes of the horses:

16 Hands	4 horses
15½ Hands	74 horses
15 Hands	138 horses
14½ Hands	83 horses
	299 horses

and the ages as:

Ages	*Number*
12 years	5
11 ,,	6
10 ,,	16
9 ,,	20
8 ,,	34

Officer of the 14th Light Dragoons, *c.*1802—an important water colour by Dighton of Lt. Fenwick, showing the extreme elaboration of details of the officers' dress of this period. It is clear from other contemporary sources that the minute details of the tracing braiding and lace decoration to cuffs and breeches differed from one officer to the next, probably because tailors had no precise patterns. Note the edging to the boots, the huge plume, the Mameluke sabre with its fancy leather and metal scabbard, the leopard turban, the cuff trefoil and the swagger cane. This plate epitomises the light cavalry 'buck' of the period.

Ages		*Number*
7	”	42
6	”	42
5	”	65
4	”	54
3	”	15
		299

A comparison with those of the Scots Greys shows that the majority of the latter were 15 hands with an equally large proportion being 15½. A similar comparison with the age of the horses confirms that the majority were five years old.

The Inspecting General found that the horses of the 10th were of good appearance and well nourished, although nine were found to be unfit for service.

Remounts

Each year the regiment sent a party to acquire new young horses. The remount party was proportionate to the number to be purchased—usually one man to two horses, exclusive of an NCO. A quartermaster and farrier were always detailed to accompany the party, and if more than 20 horses were to be purchased an officer went also.

Each man took two snaffles, two horse cloths and two surcingles; the ridden horse always had two horse cloths on it, for the greatest care was taken not to injure the young horse's back. The officer or QM had to ensure that the young horses did not travel more than 18–20 miles a day and to see that they were fed at the full allowances. Cold mashes and an ounce of nitre was given two or three times a week.

When the remounts were received from the dealer they had their numbers cut on the offside and a proper register was started. When the party arrived back at the regiment the register was handed to the adjutant, and the commanding officer then inspected the purchase. The veterinary officer accompanied him on this parade and noted particularly any mounts which he considered unsound. They were then gently bled and were given two or three doses of a mild physic.

To accord with the regulations a dragoon horse was to be between 15 hands 1in. and 15 hands 3ins., but it was often very difficult to find a complete purchase of horses of that calibre. The best horses of the purchase were selected for the sergeants and the 'rough riders'.

Sabretasche of the 14th Light Dragoons, *c*.1812—the back view of this item as used in the British Army is rarely seen.

The first lessons were gentle, by order, and never exceeded more than one hour and the remounts were only allocated to the troops when the riding-master was quite satisfied with their progress.

Sabretasche of the 15th Hussars, 1815—note the very elaborate lace edging to this example of an officer's sabretasche; the scarlet facing; and the details of the crown. Although officers' sabretasches were usually suspended on three rings, those of light cavalry troopers were often fitted with only two rings at various periods.

Representative Orders of Battle

Egypt, 1801
Finch's Brigade
12th Light Dragoons
26th Light Dragoons
11th Light Dragoons (one troop)
8th Light Dragoons (detachment with Baird from Cape)
Hompesch Light Dragoons

Walcheren, 1809
9th Light Dragoons
12th Light Dragoons
2nd Hussars KGL

Portugal, 1809
Santarem: Detachment 10th Light Dragoons
Sacavem: Two troops 14th Light Dragoons
Lisbon: Three troops 14th Light Dragoons
Tagus Forts: Three troops 14th Light Dragoons

Force to attack Isle of Ischia, 1809
20th Light Dragoons with selected Mounted Infantry

Mauritius, 1810
One troop 25th Light Dragoons

Java, 1811
22nd Light Dragoons (dismounted)

New Orleans, 1814
14th Light Dragoons (dismounted)

West Indies, 1794
Fifty men from each of 7th, 10th, 11th, 15th, 16th Light Dragoons (dismounted)

Maroon Revolt, Jamaica
Detachments 13th, 14th, 17th, 18th Light Dragoons, plus 20th LD later

Windward Islands
26th Light Dragoons

S Domingo, 1795
13th, 14th, 17th, 18th, 26th, 29th and 31st Light Dragoons, with Irving's Hussars and Hompesch Light Dragoons

South America, 1806
Officer and six men from 20th Light Dragoons with Commander Sir Hope Popham RN. Two squadrons 20th Light Dragoons and one squadron 21st Light Dragoons with Sir David Baird; the 9th Light Dragoons with Mahon. The dragoons all fought dismounted, but some managed to acquire mules.

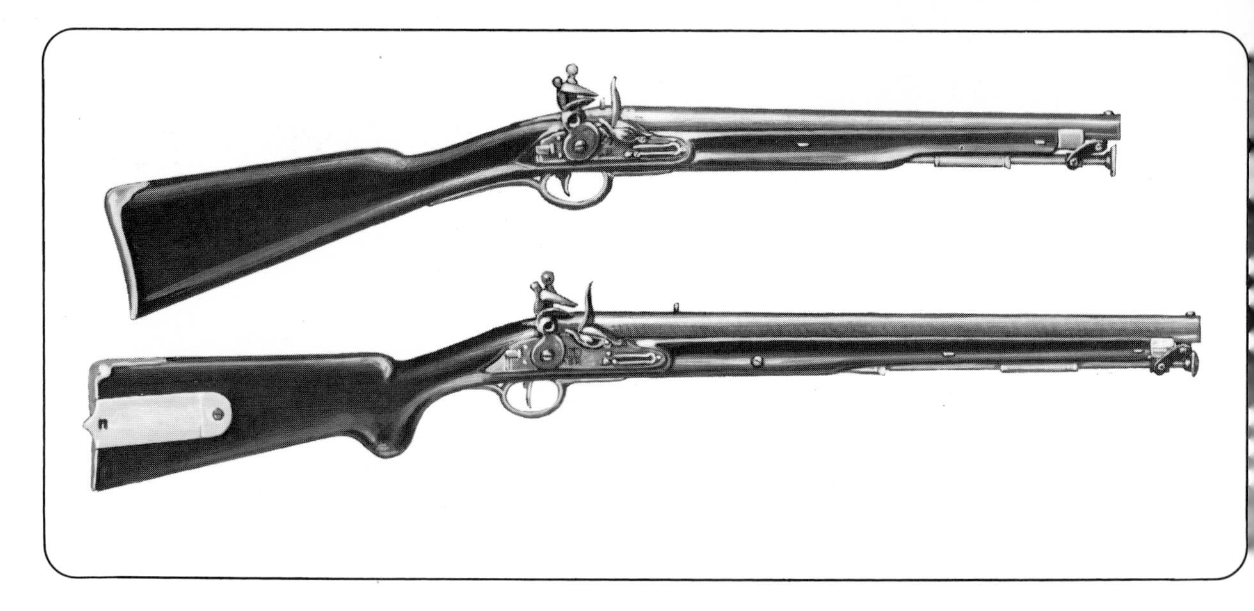

Top: light cavalry carbine (Paget) used by both Hussars and Light Dragoons during the Waterloo campaign. Bottom: Baker cavalry rifle, with pistol grip—this example is marked 'XRH' for the 10th Hussars. (Drawings by the author, as are all other tone drawings in this book unless specifically credited otherwise.)

The Light Cavalry in Spain: April 1813

	Field Strength
Long's Brigade	
13th Light Dragoons	320
Anson's Brigade	
12th Light Dragoons	
16th Light Dragoons	725
Von Alten's Brigade	
14th Light Dragoons	
1st Hussars KGL	
2nd Hussars KGL	998
Grant's Brigade	
15th Light Dragoons	
18th Hussars	512

The Light Cavalry in the Peninsula, Jan. 1814

General Officer Commanding: Lieutenant-General Sir Stapleton Cotton

Brigades were lettered 'A', 'B', and so on. Letter is given with Brigade Commander to avoid confusion:

	Field Strength
Vandeleur's Brigade (C)	
12th Light Dragoons	387
16th Light Dragoons	415
Fane's (later Vivian's) Brigade (D)	
13th Light Dragoons	348
14th Light Dragoons	427
Vivian's (later von Alten's) Brigade (E)	
18th Hussars	427
1st Hussars KGL	426
Lord Edward Somerset's Brigade (H)	
7th Hussars	513
10th Hussars	459
15th Hussars	466

The Light Cavalry Order of Battle at Waterloo, June 1815

Third Brigade:
Major-General Sir W. Dornberg

	Officers	Men
1st Light Dragoons KGL	29	462
2nd Light Dragoons KGL	29	419
23rd Light Dragoons	28	387

Fourth Brigade:
Major-General Sir John Vandeleur

	Officers	Men
11th Light Dragoons	29	390
12th Light Dragoons	20	388
16th Light Dragoons	31	393

Fifth Brigade:
Major-General Sir Colq. Grant

	Officers	Men
2nd Hussars KGL	31	564
7th Hussars	31	380
15th Hussars	26	392

Sixth Brigade:
Major-General Sir Hussey Vivian

New Land Pattern pistol, 1802, as used by both heavy and light cavalry.

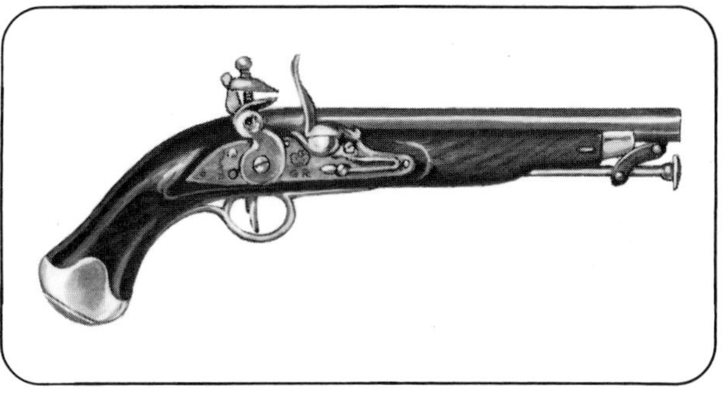

1st Hussars KGL	30	493
10th Hussars	25	390
18th Hussars	28	396
Seventh Brigade:		
Colonel Sir F. V. Arentschildt		
3rd Hussars KGL	30	622
13th Light Dragoons	29	390
Hanoverian Light Cavalry		
Prince Regent's Hussars		596
Bremen and Verden Hussars		589
Cumberland Hussars		497
Brunswick Cavalry		
Hussars		690
Lancers		232

The De Bosset Diagrams 1803

In 1803 an important information sheet, illustrating British uniforms of that time, was published by Charles Philip de Bosset, who was at that time a lieutenant in the Swiss Regiment 'De Meuron', which was in the British service. He later became a captain in the 2nd Line Battalion of the King's German Legion.

The sheet measured $24\frac{1}{2}$ins. × 19ins., and bore the following heading:

'A VIEW OF THE BRITISH ARMY / ON THE PEACE ESTABLISHMENT / IN THE YEAR 1803 / DEDICATED BY PERMISSION OF HIS ROYAL HIGHNESS THE DUKE OF YORK, COMMANDER IN CHIEF ETC ETC / BY HIS MOST HUMBLE / MOST OBEDIENT / AND / MOST DUTIFUL SERVANT / CHARLES PHILIP DE BOSSET / SWISS REGT. DE MEURON'

For each regiment there is a coloured diagram, each part painted in the colours of the various parts of the uniform. The diagram indicates the colour of the coat, its collar and cuffs, the colour and shape of any lapels or breast looping, the colour of the woven-in stripes in the lace of the private soldiers, the colour of the breeches and the colour of the officers' lace and buttons. The plate is also embellished with groups of figures to show the general form and style of the uniforms.

The Light Cavalry are shown as follows (G = gold lace, S = silver):

Regiment	*Jacket*	*Facings*	*Lace*	*Remarks*
7th LD	Blue	White	S	
8th LD	French grey	Scarlet	S	
9th LD	Blue	Pale buff	S	Buff breeches
10th LD	Blue	Yellow	S	A frame round the lace
11th LD	Blue	Pale buff	S	Buff breeches
12th LD	Blue	Pale yellow	S	
13th LD	Blue	Pale buff	G	Pale buff breeches
14th LD	Blue	Orange	S	
15th LD	Blue	Scarlet	S	
16th LD	Blue	Scarlet	S	
17th LD	Blue	White	S	
18th LD	Blue	White	S	
19th LD	French grey	Yellow	S	
20th LD	Blue	Yellow	S	
21st LD	Blue	Pale yellow	S	
22nd LD	French grey	Scarlet	S	

From various other sources we know that the following regiments were dressed as indicated:

23rd LD	Blue	Yellow*	S
24th LD	Blue	Yellow	S
25th LD	French grey	Red	S
26th LD	Blue**	Purple-blue	S
27th LD	French grey	White	S
28th LD	French grey	Yellow	S
29th LD	Blue	Pale buff	S

* One source states crimson red facings.

** Hamilton Smith shows the 26th in a French grey jacket with red facings in 1800.

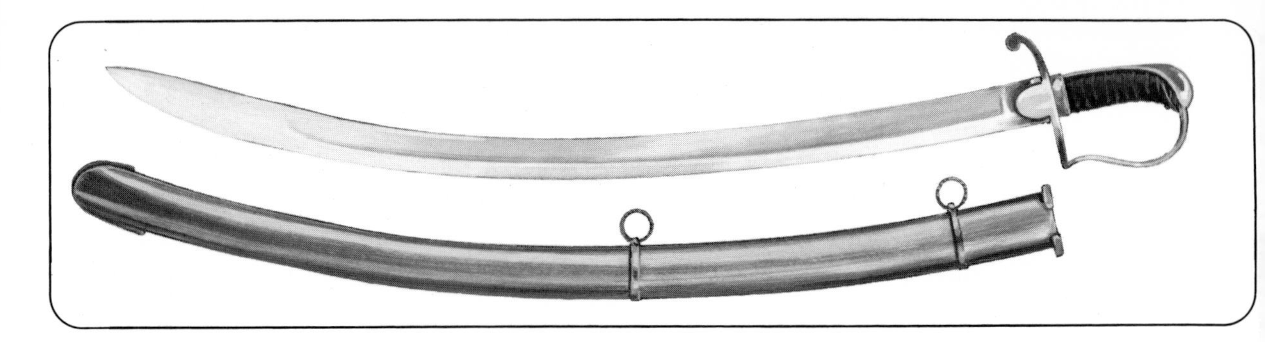

Officer's Light Cavalry sabre, 1796 pattern.

The Hamilton Smith Charts 1812

The charts provide a quick reference to the colours of the uniforms of the Light Cavalry. There are two diagrams, that for the hussar regiments showing the facing colours, the colour of the braid and buttons, the colour of the jackets (and presumably the pelisses), the barrel sashes and the breeches. The diagram for the Light Dragoons provides details of the new-style clothing which was introduced in 1812. The sections indicate the facing colours, the colour of the jackets, the colours of the worsted girdles and the buttons and officer's lace.

Hussar Regiments All have blue jackets and white breeches.

Regiment	*Facings*	*Braid and Buttons**	*Sash*
7th	White	White	White, blue barrels
10th	Scarlet	White	Crimson, yellow barrels
15th	Scarlet	White	Crimson, white barrels
18th	White	White	White, blue barrels

* Silver lace and buttons for the officers.

Light Dragoon Regiments

Regiment	*Facings*	*Lace and Buttons**	*Girdle*
8th	Red	Yellow	White, 2 blue stripes
9th	Crimson	Yellow	Yellow, 2 blue stripes
11th	Pale buff	White	Buff, 2 blue stripes
12th	Yellow	White	Yellow, 2 blue stripes
13th	Buff	Yellow	Buff, 2 blue stripes
14th	Orange	White	Orange, 2 blue stripes
16th	Scarlet	White	Scarlet, 2 blue stripes
17th	White	White	White, 2 blue stripes
19th	Yellow	Yellow	Yellow, 2 blue stripes
20th	Orange	Yellow	Orange, 2 blue stripes
21st	Pink	Yellow	Pink, 2 blue stripes
22nd	Pink	White	Pink, 2 blue stripes
23rd	Crimson	White	Crimson, 2 blue stripes
24th	Light grey	Yellow	Light grey, 2 blue stripes
25th	Light grey	White	Light grey, 2 blue stripes

* Officers' lace and buttons gold or silver.

The 21st Light Dragoons may not have adopted their distinctive pink facings as, in 1814, they were given authority to adopt black facings with silver lace. The regiment was abroad and it seems likely that the pink and gold distinctions were never brought in. Strangely, after all the fuss, gold lace was adopted in June 1815.

It seems that pink was an objectionable facing colour, for the 22nd also raised resistance to its use, and in October 1814 white facings were permitted. There is evidence that in December 1814 the men were wearing new jackets with pink facings but the officers had red facings. In July 1815 red facings were permitted, but this order was rescinded and by November 1815 white facings were ordered to be worn.

Tin helmet worn by the 8th Light Dragoons in 1796. Tin helmets were worn as replacements for the leather Tarletons in tropical climates. (D. S. V. Fosten)

King's German Legion

Regiment	*Facings*	*Braid and Buttons**	*Sash*
1st Hussars	Scarlet	Yellow	Crimson, yellow barrels
2nd Hussars	White	Yellow	Crimson, yellow barrels
3rd Hussars	Yellow	White	Crimson, yellow barrels
1st Lt. Dragoons	Crimson	Yellow	Blue, 2 crimson stripes
2nd Lt. Dragoons	Crimson	White	Blue, 2 crimson stripes

* Officers' lace and buttons gold or silver.

Hamilton Smith follows his charts of the King's German Legion with a diagram for the *Brunswick Hussars*: All black, pale blue cuffs, black braid and buttons, black breeches. Yellow sash with light blue barrels. *The Royal Staff Corps* (titled 'Staff Dragoons' by Hamilton Smith) are shown with a Light Dragoon style of clothing with scarlet jackets; blue collar, lapels and cuffs; white lace and buttons, blue girdle with two red stripes.

By 1815 there had been some amendments to the facings, lace and buttons of the Hussar regiments, and the details of the uniforms worn in that year are as follows:

The 7th (Queen's Own) Light Dragoons (Hussars)
Blue facings; gold lace and buttons for the officers and yellow for the rank and file.

The 10th (or Prince of Wales's Own) Light Dragoons (Hussars)
From 1814 the facings were blue; gold lace and buttons for the officers, yellow for the rank and file.

The 15th (or King's Own) Light Dragoons (Hussars)
Scarlet facings; silver lace and buttons for the officers and white for the rank and file.

The 18th Regiment of Light Dragoons (Hussars)
White facings; silver lace and buttons for the officers and white for the rank and file.

The Plates

A1: Trooper, 10th (Prince of Wales's Own) Light Dragoons, 1787
A2: Officer, 7th (Queen's Own) Light Dragoons, 1793
A3: Trooper, 15th (or King's) Light Dragoons, 1794

The change in colour of Light Dragoon clothing from red to blue was ordered in 1784 and the instructions consolidated by 1785. As early as 1780 there is evidence that a black leather helmet with a peak and fur crest was being utilised by the light cavalry. At first there were several variants, but by 1787 the pattern shown in A1 had been firmly established. In some reports it is called the 'helmet-cap' and a 1788 Clothing Board confirmed that it was to have a peak in front, a 'turban' to let down as a cape, a black bearskin crest, and a title plate. The turban and feather were to be of the facing colour although, by 1794, the former were mainly blue and the latter white-over-red. Regiments which had distinguishing badges were allowed to display them on the right side. This helmet became popularly known, even by the French, as the 'Tarleton', after Banastre Tarleton, the light cavalry commander of the Legion of that name, which became well known during the American War of Independence.

An order of April 1784 specifies the new blue clothing to be worn by Light Dragoons. It comprised a dark blue jacket and 'shell', an under-waistcoat and leather breeches. The collar and cuffs of the jackets were red for Royal Regiments and of a facing colour for others, and both garments were looped on the breast and edged with white cord. They were lined white, exceptions being the 11th and 13th Regiments, which had buff linings. Because the jacket and shell were the same colour the sleeves of the former are often misconstrued as being part of a single garment.

A1 is taken from a set of water colours executed by Dayes or Scott, and shows yellow linings for the 10th. Note that the artist shows a carbine swivel belt and a bayonet belt and that the long cartridge pouch is worn on a narrower belt around the waist. Shortly after 1787 the swivel and pouch belt was redesigned.

A2 shows the uniform of an officer in 1793 and is taken from two portraits of the period. An Inspection Report of eight years earlier confirms that the officers of the 7th Regiment were still not wearing the new pattern clothing, although the rank and file were. Note particularly the silver scaled wings, and the method of suspension of the sabre.

A3 is a reconstruction of the rear detail of the jacket from an actual garment which was preserved in the Zeughaus, Berlin; a drawing by Charles Hamilton Smith; and Francis Wheatley's '*The Encampment at Brighton*'. The last-named painting confirms the blue turban. In 1790 the Adjutant General had written to the commanding officer of the 15th Regiment complaining that at an earlier review the men of the regiment had red wings on their 'upper jackets' and that the deviation should be discontinued. The pouch was at this period already carried on the swivel belt.

B1: Officer, 12th (The Prince of Wales's) Light Dragoons, 1805
B2: Trooper, 11th Light Dragoons, 1798
B3: Officer, 14th (Duchess of York's Own) Light Dragoons, 1800

B1 is taken from a portrait of an officer of the 12th Regiment by Robert Dighton, which is in the collection of Her Majesty the Queen. It shows the development of the Light Dragoon uniform in the 11 years since it was formally introduced. The jacket and shell have been discarded in favour of

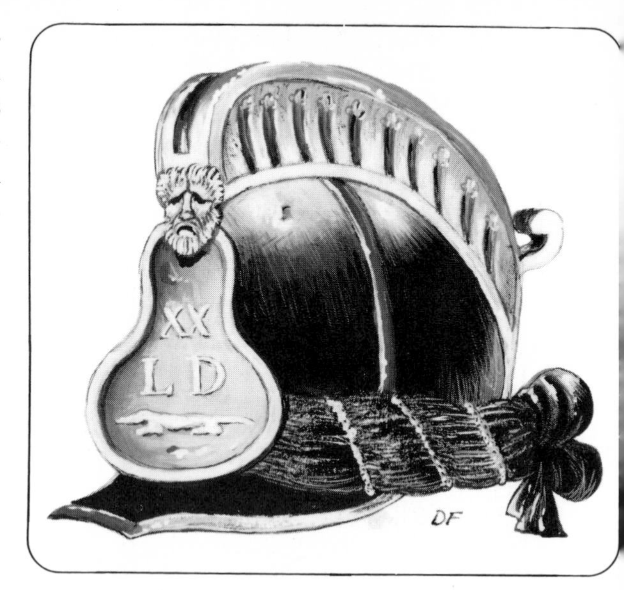

Tin helmet worn by the 20th Light Dragoons. The missing horsehair plume was probably red. (D. S. V. Fosten)

a single close-fitting garment with a more elaborate silver braid looping on the breasts. The jacket has a distinctive silver braiding above the facing-colour cuffs, and the crimson sash closely follows the pattern and style of the Light Infantry type. The regimental badge is taken from Northcote's painting of officers of the regiment receiving the blessing of the Pope when a detachment of the regiment was sent from Corsica to the Papal States, where it was quartered in the Cittia Vecchia.

B2 depicts the development of the troopers' uniform four years after its introduction. The turban is now blue and the feather white-over-red; the swivel and pouch belt have been combined and are worn over the left shoulder. The sword is the 1796 pattern light cavalry sabre with a broad and sharply curved blade approximately 33ins. long, a steel stirrup bow guard, a black grip and a steel scabbard. This weapon is said to be one of the best of British-designed swords, with an excellent balance; its weight and curved blade gave it a formidable cutting power. The 11th Regiment served in Flanders with the Duke of York, although a detachment went to the West Indies. A small group of selected NCOs went to China on escort duties, and were received with curiosity at the Imperial Court.

B3 is from a portrait of Captain Fenwick of the 14th Regiment, and provides a clear picture of

Front and rear of a private's coat of the 16th Light Dragoons, 1788, with (inset) button detail. This coat, together with the shell jacket and other articles of clothing, was preserved in the Zeughaus Museum in Berlin. Collar and cuffs were red, and lace white.

how the costume of the Light Cavalry officer was developing and becoming more elaborate. Although two troops of the regiment went to Flanders and were subsequently drafted into the 8th Light Dragoons, the remainder went to the West Indies, where they were badly mauled at San Domingo. Much reduced by casualties and disease they returned to the United Kingdom in 1797 and the following year the regiment received its title in honour of the marriage of the Duke of York to the Princess Royal of Prussia. As a result the regiment received the distinction of wearing the Prussian eagle on its appointments, and having orange facings. Part of the regiment served with Abercromby in Egypt in 1801.

C1: Officer, 22nd Light Dragoons, 1807
C2: Officer, 10th (or Prince of Wales's Own) Light Dragoons, 1805
C3: Colonel, 7th (or Queen's Own) Light Dragoons, 1798

The fourth Light Dragoon regiment to be numbered the 22nd was raised in 1802, as the 25th. As the 25th it had served at the Cape and later at Seringapatam and at Assaye. As the 22nd it subsequently served on Java and in the bloody Mahratta and Pindaree Wars, but was disbanded in 1819.

C1 is taken from Robert Dighton's portrait of Lieutenant James Jones and is dated 1807. C. C. P. Lawson showed a similar costume, but gave the figure a plain black tapered peakless cap with gold and crimson cap lines. In January 1796 the Adjutant General had ordered that the clothing of Light Dragoon regiments serving in the East Indies and the Cape should be grey rather than dark blue. There ensued a considerable correspondence between him, the Board of General Officers and the King debating the issue; but finally, in July the same year, the King decided the issue and grey clothing was approved. The Jones portrait underlines the extent to which Light Dragoon uniforms had elaborated since 1784. The cocked hats were worn for undress parades and around the cantonments and on other occasions when the helmet was unsuitable. Note particularly the extraordinary decoration on the sleeves and breeches, the black sword belt and slings, the Mameluke sabre, and the curious all-crimson but barrelled sash.

C2 is similarly derived from Dighton and shows an officer of the 10th. From 1796 this regiment had the Prince of Wales as its colonel, and under his patronage exceeded most in the extravagance of its dress. The Prince took a great interest in the regiment's economy and uniform. His Wardrobe Accounts reveal the resources which he was prepared to invest in regimental dress. Even so the Dighton painting shows an extraordinary elaboration. The cap—known in France as the

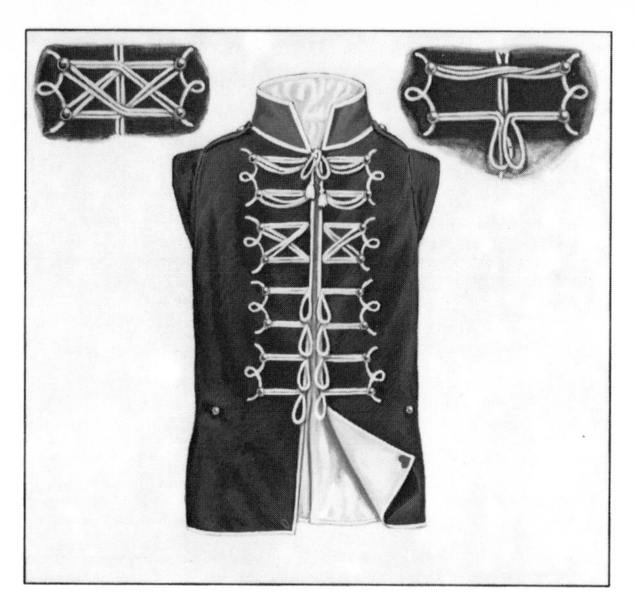

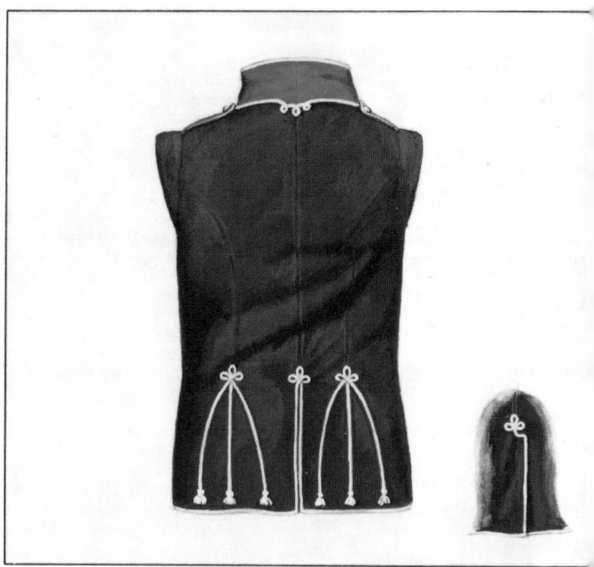

Front and rear of a shell jacket of the 16th Light Dragoons. The method of fastening across the breast is shown in many contemporary illustrations. Although the majority are as the sketch at top left, there are many variations. It is interesting to note that C. C. P. Lawson and P. W. Reynolds, both meticulous in research, sketched the same article with slight variations: Lawson shows the rear vent as inset on the right, and omits shoulder straps.

'mirliton' and in Prussia as the 'Flügelmutze'—was perhaps originally derived from the Turkish janissary headdress, but was certainly copied, in the British Army, from the headdress used by some emigré regiments. The yellow boots were considered very much 'à la mode' by young 'bucks' of the period. In the Creevey Papers, Volume 1, we find '. . . the Officers of the Prince's Regiment [i.e. George, Prince of Wales] had all dined with him and looked very ornamental monkeys in red breeches with gold fringes and yellow boots . . .'

In contrast Hoppner's portrait of Colonel Lord Paget of the 7th Regiment—see *C3*—shows a similarly handsome and distinctive but less elaborate uniform, which is only a slightly modified version of the original blue under-jacket. Note the very plain cuffs. In the 7th Regiment this jacket was worn until about 1805, when extra braid and a more elaborate cuff was introduced. Colonel Lord Paget is shown off duty with his jacket unbuttoned to reveal the handsome silver-braided white waistcoat, and wearing silver-laced dark blue breeches and plain boots.

D1: Trooper, 7th (Queen's Own) Light Dragoons, Hussars, 1808

D2: Sergeant Major, 15th (or King's) Light Dragoons, Hussars, 1808

D3: Officer, 10th (or Prince of Wales's) Hussars, 1808

As early as 1801 there had been indications that certain Light Dragoon regiments were to be converted to Hussars. Caps such as the 'mirliton' were being introduced, along with fur-trimmed pelisses and sabretashes; and it is known that in the 7th the tapering 'mirliton' cap had been in use for several years.

In 1806–7 the 7th, 10th and 15th Light Dragoons were formally converted to Hussars although, even then, the ever-conservative gentlemen at Horse Guards were reluctant to overstate the change and in two cases only added the title 'Hussar' as an afterthought.

All three figures on this plate have been based on splendid water colour drawings by Robert Dighton junior, now in the collection of the Duke of Brunswick and carefully recorded by W. Y. Carman. They were formerly in the collection of Ernst August, King of Hanover, who, as Duke of Cumberland, had been colonel of the 15th Hussars from 1801 to 1826.

All three wear the tall muff-like fur cap. This was not a popular headdress and proved ineffectual as a cap for campaigning. It was a fur cylinder, with the red bag (almost a bonnet) rising up the full height of it to emerge freely from the op; it could fall on either side. Note that the cap f the officer of the 10th is grey, and this is con-rmed in two other sources: the portrait of Colonel Quentin and the group of Light Cavalry fficers dated 1808, both in the Royal Collection t Windsor. Officers of the 15th had black fur caps.

D1 portrays a trooper of the 7th, and points to note are the stubby tuft in the cap, the blue stripe n the collar and the colours of the barrelled sash. C. Hamilton-Smith shows a light blue and dark blue sash in his print dated 1813. In the 7th the officers' pelisses had brown fur, but it was white or the rank and file.

D2 is an NCO and probably represents a ergeant major, although in 1801 all NCOs of the 5th Hussars had been granted crowns over their hevrons. However, as the sash seems definitely rimson with gold barrels this seems to confirm he higher rank. It is unclear whether the artist ntended to represent white or silver lace in this ase. Note that the pelisse has brown fur.

D3 is from the same source and portrays an fficer of the 10th Hussars in the distinctive grey ur cap, which is reflected in the similarly grey fur f the pelisse. While officers of this regiment had ed bags to their caps, the other ranks had brown ur caps with yellow bags. The Prince of Wales's Wardrobe Accounts suggest that a cap with a ellow bag was proposed for the officers but never adopted. Note that although the pouch belt is yellow faced silver, the sword belt is a dull red worked with silver embroidery.

In 1808 all three of these regiments were in the Peninsula, no doubt wearing this uniform.

E1: Officer, 13th Light Dragoons, 1813
E2: Sergeant, 20th Light Dragoons, 1812
E3: Trooper, 22nd Light Dragoons, 1812

Hussar regiments had experimented with various styles of shakos as substitutes for the cumbersome fur caps between 1809 and 1812, but a shako was not properly authorised for general service in Light Dragoon regiments until August 1811, and even then was not brought into universal use until the following year.

The new cap was slightly wider at the top than the bottom, had a plain peak, a deep turned-up reinforcement at the rear, a cut-feather plume for officers and a worsted tuft for other ranks. Officers had a wide gold or silver lace band around the top edge, a corded gold or silver boss in front connected by a loop and button to a tracing braid 'cartwheel' ornament on the front, and gilded or silvered chin scales. For other ranks the ornaments were white or yellow and the chin scales white metal or brass. The central circlet was also white or yellow and had, as a central ornament, a number or device for certain regiments. For example, the 9th LD, had 'IX', while the 13th LD had an eight-pointed star. The 17th LD had the skull and crossed bones instead of

Front and rear of a trooper's jacket of the 13th Light Dragoons, of the type introduced in 1812. The facings are buff. Note the 'waterfall' ornament at the rear of the waist.

Front and rear of an officer's laced jacket of the 11th Light Dragoons, 1811; facings are buff and the lace is silver. Although it is not shown in this illustration it is quite common to find a hook sewn centrally into the bottom edge of the rear of the jacket; this held the barrelled sash in position and stopped jacket and sash becoming separated.

the circle, the 11th LD had the 'XI' on the boss but a Sphinx in the circle itself; and in the 12th LD the Sphinx was displayed in the circle without any number. Cap lines were crimson mixed with gold for the officers, and white or yellow for the other ranks. Wellington is said to have disliked the new cap, primarily because it gave rise to the British Light Dragoons being mistaken for the French.

The old-style braided jacket was also now discarded in favour of a dark blue jacket with collar, plastron, cuffs, turnbacks, and rear seam piping in a variety of facing colours. Epaulettes were worn on both shoulders, and there were buttons plus a 'waterfall' of lace at the waist at the rear. A girdle was introduced, gold and crimson for the officers but dark blue with facing colour stripes for the other ranks. Officers and men wore white breeches and hussar-pattern boots for dress parades on the home establishment, but blue-grey overalls with two stripes down the outward seams on active service and for undress duties. Pictorial evidence suggests several regiments wore red stripes on these overalls irrespective of the facing colour, while other pictures suggest that they were of the facing colour.

In the West and East Indies and other tropical stations the new clothing was worn, but with special brown shakos.

The 21st and 22nd Light Dragoons were given fancy pink facings with, respectively, gold and silver buttons and lace. In 1814 permission was given for the 21st to change to black facings, and in June 1815 they were given gold lace. Similarly the 22nd were granted white facings in October 1814, but by the end of that year are recorded as still having pink, although their officers had adopted red. In July 1815 red facings were properly authorised, but in November the same year white facings were introduced.

F: King's German Legion
F1: Officer, 1st Regiment of Light Dragoons, 1815
F2: Officer, 2nd Regiment of Hussars, 1815
F3: Trooper, 2nd Light Dragoon Regiment, 1815

Although it is always referred to as the King's German Legion its units were never employed as a composite and separate force in the field. When on active service the various elements were allocated to British formations, and its higher command structure was only responsible for recruiting, supplies, and internal promotions and so forth. However, the Legion had its own Adjutant General, and staff work in the field was dealt with by KGL Brigade Majors in the usual way. Wellington had a KGL aide-de-camp on his personal staff. In the Legion the Cavalry ranked third in precedence, the senior arms being the Engineers and the Artillery. The majority of its fficers were from units of the disbanded Hano-erian Army and were mainly drawn from the obility and landed gentry, many coming from he great families, such as the von der Deckens and on Linsingens. The Colonel-in-Chief was the rinz Adolphus Friedrich, Duke of Cambridge, vho ranked as a lieutenant-general and was the ormer Colonel-in-Chief of the Hanoverian Guard Regiment.

Each of the cavalry regiments had the same stablishment. The RHQ comprised the colonel r lieutenant-colonel, a major, an adjutant, a paymaster, a surgeon, two assistant surgeons a eterinary officer, a sergeant major, a paymaster ergeant, a saddler, a riding master and a farrier. The regiments were formed in four squadrons, ach of two troops; but in 1812 an additional quadron was added to each establishment, only o be reduced again just before the opening of the Waterloo campaign. Each of the troops had the following establishment: 1 captain, 1 lieutenant, 1 cornet, 1 quartermaster sergeant, 4 sergeants, 4 corporals, 1 trumpeter and 76 troopers.

(One trumpeter was designated 'Stabstrompeter', although there was no establishment.)

Remount training and equitation training within each troop was controlled by one officer (usually the lieutenant), with two corporals who were given extra pay for the duty.

In 1813 the two Heavy Dragoon regiments were converted to Light Dragoon status because of the deficiency of light cavalry in the field; but the new clothing was slow to arrive, and the

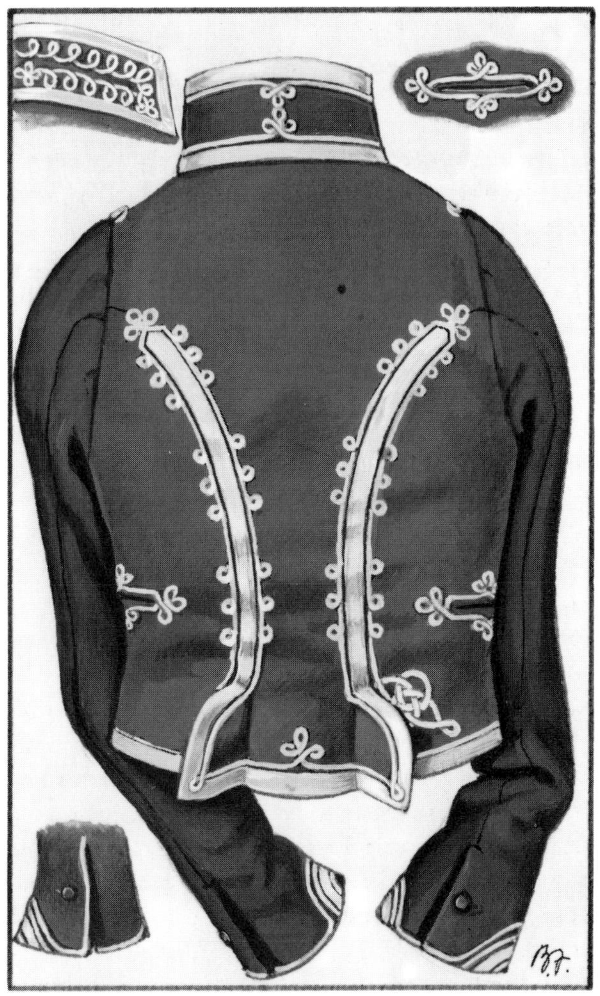

Front and rear of a trooper's jacket of the 7th Hussars, 1815. The original jacket was, until at least 1957, in the Christchurch Museum, Ipswich. It is attributed to Private Simmonds and was at one time in the Cotton Museum at Waterloo. A second jacket in the same museum shows similar details apart from the collar and the rear of the cuff—these are shown inset.

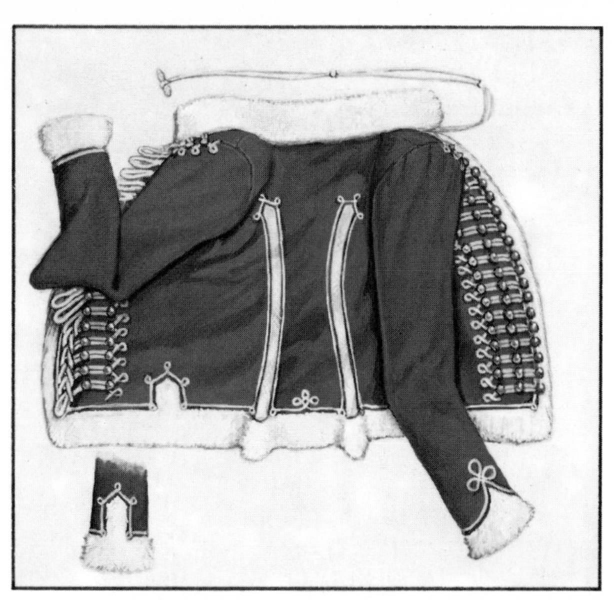

A trooper's pelisse of the 18th Hussars. C. Hamilton Smith and Herbert Knötel show similar versions differing only in cuff detail: Hamilton Smith's version is shown inset.

Inspection Reports reveal that as late as 1814–15 Light Dragoons in Belgium were still dressed in a mixture of the old red and new blue clothing.

The KGL Cavalry had a high reputation, which is reflected in their battle-honours for the Peninsular and Waterloo campaigns. These honours were subsequently proudly passed to the relevant Hanoverian regiments of the First Reich.

G1: Corporal, Staff Corps of Cavalry, 1814
G2: Officer, 18th Hussars, 1814
G3: Officer, 7th (Queen's Own) Light Dragoons, Hussars, 1810

G3 depicts an officer of the 7th Hussars in 1810 wearing the curious peakless shako, which was used by the regiment in substitution for the tall fur cap. The plume remains the traditional white-over-red, but the cap is brown with elaborate gold lace decoration and cap lines. This type of peakless cap was adopted by Hussar regiments and worn for a short period, but in the 7th, at least, a more traditional shape with a peak and covered in blue cloth was introduced in 1811. In the field the regiment wore grey overalls with two white stripes up the outward seams, white pouch belts, plain black pouches, and plain black sabretasches. Note that the hussar pelisse was now a very ostentatious garment, elaborately decorated with silver lace and braid in much the same style as the jackets. The regiments had blue shabraques with white vandyking around the edges, and adopted sheep skin covers for the saddles.

G2 is based on a miniature portrait of Captain Dolbel of the 18th Hussars dated 1814, and shows in detail the development of the style of braiding for the hussar jacket. Note the very small white cuff and the heavy silver braiding up the sleeve, also the odd method of tying the barrel sash. The captain's method of wearing his jacket open to reveal the equally heavily decorated white waistcoat, and the method of tying the barrelled sash around the waist over it but under the jacket confirms that this was the appropriate informal dress utilised by officers when not on duty, around the barracks and cantonments, for drawing-room and other social functions, and for undress duties in hot weather.

G1 portrays the uniform of the Staff Corps of Cavalry, which was formed in 1813 as a replacement for the very hard-pressed department of the Provost Marshal. Four troops were raised under the command of Sir George Scovell, and attached to the Adjutant General's Department. The duties of the Corps were to patrol the camps and cantonments, to pick up and deal with stragglers, to protect the local populace from any excesses by the troops, and, on occasion, to provide orderlies for the staff. It had a reputation for being tough but generally fair, and was very highly thought of by Wellington. After the Peninsular Wars it was disbanded, but troops were raised for the subsequent Waterloo campaign.

All the other ranks were either reliable senior NCOs or long-service troopers with clean records. Two troops were formed from regiments on the Home Establishment and sent out to the Peninsula and another two were raised from regiments in the field. Both officers and other ranks received extra pay for their duties.

The officers' uniform had silver lace, epaulettes and buttons and they had crimson and gold cap lines.

H1: Trooper, 7th (or Queen's Own) Light Dragoons, Hussars, 1815
H2: Trooper, 18th Hussars 1815
H3: Officer, 10th (or Prince of Wales's Own) Hussars, 1815

In 1811 the white-faced jackets of the 7th Hussars

vere replaced by plain blue clothing with gold ace and buttons (yellow for other ranks). At first hakos were worn with the new uniforms but, by 1815, brown fur caps with red bags, brass chin cales, and button-colour cap l·nes had been e-introduced, although they still proved un-popular as campaign headdress. Lieutenant Uniacke wrote that an enemy '. . . cut my fur cap down the centre and gave me a cut on the orehead . . .' The trooper's jacket is copied from n actual specimen worn by Trooper Simmonds luring the Waterloo campaign.

The 18th Hussars had worn the same uniform ince their conversion in 1807; and in 1813 they paraded at Luz in the Peninsula, and a note was aken of their clothing on that day. All ranks were wearing the fur cap, jackets, slung pelisses, white oreeches and hussar boots. On occasion the men nad to parade in their stable trousers as the over-alls were deficient. The Regimental Inspection Report for September 1815 confirms that the rank and file had jackets, pelisses, breeches, fur caps and gloves. Although the report specifies dress and undress jackets it does not mention overalls, which is curious. Dighton portrays the fur cap as being a squatter but broader headdress than the old style 'muff' cap and it has a bright blue bag. His painting indicates grey overalls with red stripes up the outward seams for the rank and file.

According to Commissariat Schaumann the 10th Hussars were wearing red shakos as early as the spring of 1812, and it is known that they had already been wearing black shakos, probably of the peakless pattern, since 1809. However, for Waterloo there are several contemporary sources which confirm that the red shako was worn, and most of the detail can be taken from two fine paintings by Dighton of a charge of the Hussars led by the Marquis of Uxbridge. The caps have plain black peaks and black leather reinforcement at the rear, and a band of (apparently) white braid (silver for the officers) around the upper edge. Officers' caps have a row of interlocking gold rings beneath the upper lace band, which may be an early method of distinguishing rank on the head-dress. In 1814 the 10th had been ordered to take into use blue-faced jackets with gold lace and buttons (yellow for other ranks) replacing the red-faced-blue jackets with silver lace which were then in use. A regimental tradition was maintained in the flat lace frame around the braid looping on both jackets and pelisses. It is curious that Dighton takes pains to show that the 10th, and they alone, had queues in 1815. This regimental peculiarity is not shown in other contemporary prints and could be thought to be an error, except that Dighton has repeated the detail in another painting of the regiment in Spain—which, incidentally, shows the peakless shako.

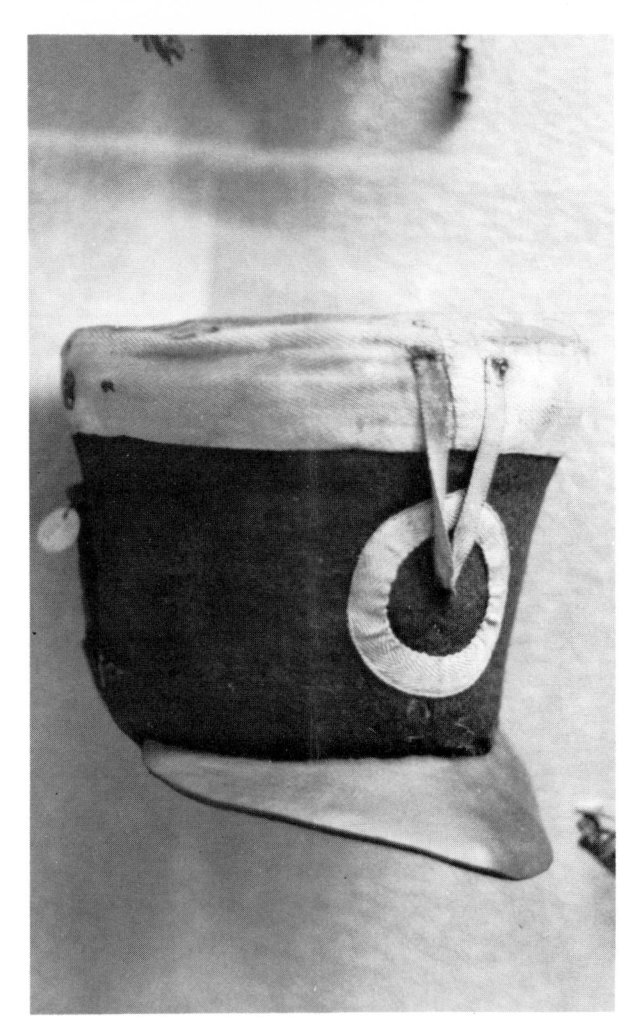

An example of the Light Dragoon shako introduced in 1812. It is interesting to note the roundel of coarse lace sewn directly on to the felt and, just visible here, the up-turned false peak or neck cover at the back of this piece. (Photo: Paul Richardson)

Notes sur les planches en couleur

A Les uniformes des Dragons Légers passèrent du rouge au bleu en 1784–85. Notez la 'carapace' sans manches pardeussus la veste à manches longues. Le casque de cuir ètait utilisé dès 1780. On l'appellait le 'Tarleton' d'après un officer célèbre de la guerre américaine d'Indépendance. Dans **A1** la giberne est portée à taille et il y a des courroies différentes pour la carabine et la bayonette. **A3** La giberne est ici passée sur la courroie de carabine.

B Peu d'années après son dotation, la cavalerie légère a déjà un uniforme très élaboré. La veste et la 'carapace' ont maintenant été remplacées par un seul vêtement avec des boucles de galons sur la poitrine. Le sabre de la cavalerie légère de 1796 (**B2**) a la réputation d'être la meilleure arme blanche militaire anglaise qui aie jamais été produite—son poids et la courbe incisive de la lame lui domnentune grande puissance de coupe, et elle était très bien équilibrée.

C1 Uniforme extraordinairement compliqué, reproduit d'après un portrait, destiné aux officiers servant sous les tropiques, où le gris remplaçait le bleu habituel. Le bicorne ètait couramment porté par les officiers en permission. **C2** La coiffure 'mirliton' est originale. L'armée anglaise l'avait copiée des régiments français émigrés. Le prince Régent en était le colonel et les uniformes élaborés témoignent de son intérêt. **C3** Le colonel Lord Paget dans un uniforme plus sobre, porté en permission.

D En 1806–07 les 7ème, 19ème et 15ème Régiments des Dragons Légers devinrent hussards officiellement. Ces illustrations ont été faites d'après des aquarelles de Robert Dighton le jeune, appartenant au duc de Brunswick. La haute casquette de fourrure n'ètait ni populaire ni pratique. La coiffure de fourrure grise ètait particulière aux officiers du 10ème. **D2** C'est probablement un sergent-major, mais tous les sous-officiers du régiment portaient une couronne couronne au-dessus de leurs chevrons.

E Un nouveau shako fut distribué à la cavalerie légère en 1811–12 et la veste à galons fut remplacée par un autre modèle revers dans les couleurs du régiment, à hauteur de poitrine. Une version marron du shako était portée aux tropiques. A cette qpoque, les couleurs des revers de plusieurs régiments changèrent aussi de façon complexe. Voir la liste des details dans le texte principal.

F La magnifique cavalerie légère de la King's German Legion, composée d'une majorité de Hanovriens, avait l'enviable réputation d'être la plus professionnelle des cavaliers de Wellington. Les régiments de dragons lourds de la King's German Legion furent convertis en dragons légere en 1813 mais leur uniforme manquait encore d'homogenéité deux ans plus tard: le rouge d'avant 1813 se trouvait encore parmi les nouveaux uniformes blaus.

G1 Cette unité de quatre pelotons était formée d'hommes sûrs venus d'autres regiments. Ils servaient de police militaire et d'ordonnances auprès des officiers supérieurs. Ils avaient une excellente réputation. **G2** Uniforme de congé, avec l'écharpe portée directement sur un gilet galonné visible sous une veste ouverte. **G3** Ce shako sans visière était une des alternatives utilisées par le régiment à la place de la haute coiffure cylindrique en fourrure.

H1 Après avoir porté des shakos pendant un certain temps, ce régiment était revenu en 1815 à la peu populaire coiffure de fourrure. **H2** Dighton représente le régiment avec une version plus ramassée de la coiffure en fourrure. **H3** L'usage de cet élégant shako rouge remonte à 1812, comme cela a été confirmé à plusieurs reprises. Des informations contradictoires suggèrent que cette unité était la seule à porter les cheveux nattés.

Farbtafeln

A Die Uniform der leichten Dragoner wechselte in den Jahren 1784–85 vor rot zu blau. Bemerke die Benutzung einer armellosen 'Hülle' über einer langärmeligen Jacke. Der Lederhelm wurde schon so frühzeitig wie 1780 benutzt; er war als der 'Tarleton' bekannt, nach einem berühmten Offizier des amerikanischen Unabhängigkeitskrieges. In **A1** ist der Patronenbeutel an der Taille getragen und es sind zwei verschiedene Gürtel, der Karabiner- und der Bajonettgürtel; **A3** zeigt die spätere Anordnung mit dem Beutel vom Karabinergürtel hängend.

B Bemerke die sorgfältige Ausarbeitung der Leichten Kavallerieuniform zu diesem Datum, ein paar Jahre nach ihrer Einführung. Die getrennte Jacke und 'Hülle' sind nunmehr durch ein einziges Kleidungsstück mit ausführlich ausgearbeiteten Tressenschlingen auf der Brust ersetzt. Man sagt, dass der Leicht Kavalleriesäbel (B2) einer der besten britischen Militarsäbel jeglicher Periode sein soll—sein Gewicht und die scharf gebogene Form gaben ihm gross Schnittkraft und seine Balance war ausgezeichnet.

C1 Aussergewöhnlich sorgfältig ausgeführte Offiziersuniform, einem Porträt entnommen; das graue Tuch ersetzte das blaue in den Truppen, die in der Tropen dienten. Der Zweispitz war normale dienstfreie Kleidung für Offiziere **C2** Die mirliton Mütze ist bemerkenswert; es war in der britischen Armee von Auswandererregimentern kopiert. Der Prinzregent war der Oberst dieser Einheit und die sorgfältig ausgearbeitete Uniform reflektiert sein Interesse **C3** Oberst Lord Paget in Dienstfrei-Kleidung von viel zurückhaltender Art.

D 1806–07 wurden die 7., 10. und 15. Leichten Dragoner offiziell Husaren diese Bilder wurden von Aquarellen von Robert Dighton dem Jüngeren nunmehr in der Sammlung des Herzogs von Brunswick, entnommen. Die hohe Pelzmütze war unpopulär und unpraktisch. Die graue Pelzmütze war eine Eigenart der Offiziere des 10.. **D2** ist wahrscheinlich ein Hauptfeldwebel bemerke jedoch, dass alle Unteroffiziere dieses Regiments das Kronenmotiv über ihren Rangwinkeln trugen.

E Ein neues zugespitztes Schako wurde für die Leichten Dragoner im Jahr 1811–12 eingeführt und die mit Tressen verzierte Jacke wurde durch ein Kleidungsstück mit zurückgeschlagenen Aufschlägen aus Regimentsbesatzfarben auf der Brust ersetzt. Eine braune Abart des Schakos wurde in den Tropen getragen. Zu dieser Zeit wechselten die Besatzfarben verschiedener Regimenter in komplizierter Weise—Einzelheiten sind im Haupttext aufgeführt.

F Die prachtvolle Leichte Kavallerie der King's German Legion, weit und breit gepriesen als die Bestausgebildetsten in der Armee Wellingtons, waren hauptsächlich Hanoveraner. Im Jahr 1813 wurden die zwei KGL Schweren Dragonerregimenter in Leichte Dragoner umgewandelt, jedoch fehlte im Jahr 1815 immer noch eine einheitliche Linie in ihrer Kleidung, es wurde eine Mischung von alter roter und neuer blauer Kleidung getragen.

G1 Diese Einheit, aus vier Zügen bestehend, wurde von zuverlässigen Männern aus anderen Regimentern gebildet; sie stellten eine Militärpolizeitruppe und Ordonanzen für Stabsoffiziere dar, die sich hohes Ansehen verdienten. **G2** Freizeituniform, mit der Schärpe direkt über einer mit Litzen verzierten Weste, die von der offenen Jacke freigegeben wurde. **G3** Eine der Alternativen der unpopulären hohen zilinderförmigen Pelzmütze war das Schako ohne Schirm dieses Regiments.

H1 Nach einer Zeit, in der Schakos getragen wurden, ist dieses Regiment beim Jahr 1815 wieder zu den unpopulären Pelzmützen zurückgekehrt. **H2** Dighton zeigt das Regiment in einer niedrigeren, jedoch breiteren Abart der Pelzmütze. **H3** Verschiedene Quellen bestätigen das Tragen des feinen roten Schakos in dieser Einheit schon von 1812 an. Gegensprüchliche Beweise deuten darauf hin, dass allein diese Einheit das Haar im Zopf getragen haben mag.